# HOME FOR GOOD

A Journey of Service, Love, & Faith

*By*
PATRICK L. PLUMMER

HOME FOR GOOD
Copyright © 2025 by PATRICK L. PLUMMER

All rights reserved. No part of this book may be reproduced or transmitted in any form or by any means without written permission from the author.

ISBN 979-8-9939507-0-9

Printed in USA by 48HrBooks (www.48HrBooks.com)

# DEDICATION

*For my wife, Taska —*

*Your strength carried our family through every storm,*

*your love brought me home,*

*and your faith shaped the man I became.*

*For my children —*

*You were my reason on the battlefield*

*and my peace when I returned.*

*For Jehovah —*

*who guided my steps long before I knew His name.*

*This book is for you.*

# CONTENTS

**DEDICATION** ............................................................. III

**ACKNOWLEDGEMENTS** ........................................ XI

**PREFACE** .................................................................. XIII

    THE DUAL JOURNEY ............................................................ XIII

    A NOTE TO THE READER ..................................................... XIV

**PROLOGUE** ................................................................. XV

**EARLY LIFE – JAMAICAN ROOTS & A NEW BEGINNING** ............................................................... 19

    GROWING UP JAMAICAN IN AMERICA ................................ 19

    THE BAND ROOM — THE MOMENT I REALIZED I WAS DIFFERENT 20

    SUNDAY BREAKFASTS & SCHOOL-DAY PORRIDGE — THE FLAVORS OF MY CHILDHOOD ............................................................ 20

    THE STRENGTH & HEART OF MY PARENTS ....................... 21

    THE "BARREL" TRADITION — GIVING BACK TO JAMAICA .... 21

    EXCELLENCE WAS THE STANDARD — NO C'S ALLOWED .... 21

    COLLEGE — GOOD GRADES, BUT NO REAL PURPOSE ....... 22

    THE DAY THAT CHANGED EVERYTHING ............................ 22

    LEAVING HOME FOR THE FIRST TIME ................................. 23

**GERMANY – MY FIRST DUTY STATION, MY FIRST REAL TEST** ................................................................. 25

    A YOUNG SOLDIER IN A NEW WORLD ............................... 25

    CULTURE SHOCK IN EVERY DIRECTION ............................. 26

    THE BOND OF SOLDIERS AWAY FROM HOME .................... 26

Checkpoint Charlie — The Unforgettable Crossing .......... 26
Training Under Russian Watch ............................................... 27
Learning Responsibility the Hard Way ................................ 27
Choosing to Become a 91B Combat Medic ............................ 28
The Training Accident — The Day I Learned What "Helpless" Really Means ........................................................... 29
The Pressure of Being "Doc" .................................................... 29
The Friendships That Stayed With Me ................................... 29
Germany — The Foundation for Everything to Come ......... 30

## MEETING TASKA — THE MOMENT EVERYTHING CHANGED .................................................................................. 31

Fort Polk — A New Challenge and a New Awareness ......... 31
A New Chapter Opening Without Me Realizing It ............... 32
The Car Ride That Changed My Life ....................................... 32
A Connection That Became Part of My Every Day ............. 33
The Two-Day Silence — When I Realized I Couldn't Live Without Her ................................................................................ 33
Two Weeks Later — A Realization I Couldn't Ignore ......... 34
The Proposal Test — A Choice That Defined Everything .. 34
How I Knew She Was the One .................................................. 35

## STARTING OUR LIFE TOGETHER — YOUNG, MARRIED, AND LEARNING AS WE WENT ................... 37

Our First Home — Humble, Colorful, and Ours .................. 37
Learning How to Be Married .................................................... 38
Money Was Tight, but Love Was Strong ................................ 38
A Young Soldier, A Young Wife — Two Lives, One Mission 39
Learning to Be a Husband ......................................................... 39
Building a Partnership, Not Roles ........................................... 39
Laughing, Learning, and Loving Through the Struggles .. 40
Preparing for the Next Chapter — Becoming Parents ....... 40

## BECOMING A FATHER — THE BEGINNING OF A NEW LEGACY ........................................................................... 41

HEARING THE WORDS THAT CHANGED MY LIFE.................................. 41
THE BIRTH OF OUR FIRST CHILD: A NEW KIND OF LOVE IN TEXAS 42
A NECESSARY EVOLUTION: FROM MEDIC TO MAINTAINER ............ 42
TWO MORE BLESSINGS IN COLORADO ............................................... 43
THE WEIGHT OF RESPONSIBILITY — AND THE STRENGTH IT BUILT IN ME .................................................................................................................. 44
WATCHING THEM GROW — THE JOY THAT MADE EVERY SACRIFICE WORTH IT ...................................................................................................... 44
THE FOUNDATION OF MY FUTURE — BUILT ON LOVE .................... 45

## CLIMBING THE ENLISTED RANKS — BOARDS, BOOKS, AND BREAKTHROUGHS ............................................................. 47

A SOLDIER WITH SOMETHING TO PROVE ........................................... 47
STUDYING FOR THE BOARD — MY LATE-NIGHT CLASSROOM ....... 48
TASKA — MY UNOFFICIAL DRILL SERGEANT .................................... 48
THE FIRST BOARD — NERVES AND NCOS .......................................... 48
PROMOTION TO SERGEANT ..................................................................... 49
LEADERSHIP — GROWING INTO THE BOOTS I WORE ....................... 49
THE DRIVE TO KEEP CLIMBING .............................................................. 50

## KOREA — A YEAR AWAY FROM MY FAMILY ........... 51

LEAVING EVERYTHING THAT MATTERED ............................................ 51
ARRIVING IN KOREA — COLD, BUSY, AND LONELY ........................ 52
TASKA — CARRYING THE WEIGHT AT HOME .................................... 53
THE STAFF SERGEANT BOARD — PRESSURE AND PURPOSE ........... 53
DAILY LIFE IN KOREA — A ROUTINE OF DUTY AND DISCIPLINE ... 54
THE EMOTIONAL DISTANCE — AND THE LOVE THAT SURVIVED IT 54
RETURNING HOME — A MAN CHANGED BY DISTANCE ................... 55

## RECRUITING DUTY — PRESSURE, PURPOSE, AND CHANGING LIVES ...................................................................... 57

A WARRANT OFFICER PATH DEFERRED .............................................. 57
A WORLD AWAY FROM THE ARMY I KNEW ........................................ 58
THE PRESSURE — NUMBERS, NUMBERS, NUMBERS ......................... 58

HELPING YOUNG PEOPLE — FINDING PURPOSE IN THE HARD DAYS ............................................................................... 59
PARENTS WHO THANKED ME .............................................. 59
THE HARDEST PART — WHEN THEY DIDN'T MAKE IT ................. 60
LONG HOURS, LONG DRIVES, LONG DAYS ............................. 60
TASKA'S STRENGTH DURING THIS CHAPTER ............................ 60
MOMENTS THAT STAY WITH ME .......................................... 61
RECRUITING DUTY PREPARED ME FOR SOMETHING BIGGER ......... 61

## THE WARRANT OFFICER JOURNEY — STEPPING INTO A NEW LEVEL OF LEADERSHIP ........................ 63

THE DECISION — A QUIET VOICE PUSHING ME FORWARD ............ 63
TASKA — THE VOICE THAT PUSHED ME FORWARD ..................... 64
APPLYING — THE WEIGHT OF THE UNKNOWN ............................ 64
WARRANT OFFICER SCHOOL — HUMBLING AND TRANSFORMATIONAL ............................................................ 64
BECOMING A 140A — A NEW WORLD OF RESPONSIBILITY ........... 65
LEADERSHIP AT A HIGHER LEVEL ........................................... 66
THE HUMILITY BEHIND THE RANK .......................................... 66
TASKA'S ROLE — THE QUIET STRENGTH BEHIND THE ACHIEVEMENT ...................................................................... 66
A NEW IDENTITY, A NEW CHAPTER, A NEW BEGINNING ............... 67

## KUWAIT — THE DEPLOYMENT BEFORE 9/11 ............ 69

LEAVING HOME AGAIN — THE WEIGHT HITS HARDER THE SECOND TIME ................................................................................. 69
ARRIVING IN KUWAIT — HEAT, SAND, AND A DIFFERENT KIND OF MISSION ............................................................................ 70
A WARRANT OFFICER IN A REAL-WORLD MISSION ..................... 70
THE RHYTHM OF DEPLOYMENT .............................................. 71
THEN... EVERYTHING CHANGED - — SEPTEMBER 11, 2001 .......... 71
THE BARRACKS THAT NIGHT — QUIET, HEAVY, UNITED ............... 71
A NEW LEVEL OF READINESS ................................................ 72
A DEPLOYMENT THAT BECAME A TURNING POINT ..................... 72

## IRAQ — WAR, LOSS, AND THE WEIGHT A SOLDIER CARRIES .................................................................................. 75

THE IMMEDIATE REALITY: BIAP ARRIVAL ....................... 75
CROSSING INTO IRAQ — A DIFFERENT KIND OF SILENCE .............. 76
IEDS — THE INVISIBLE ENEMY ....................................................... 76
THE NIGHT THE RADIO BROKE ........................................................ 77
THE ETHICAL BURDEN: LOOKING DOWN THE BARREL .................... 77
LOSING SOLDIERS — A PAIN THAT NEVER LEAVES ....................... 78
THE SOLDIER'S MASK — LEAD ON THE OUTSIDE, BREAK ON THE INSIDE .................................................................................................... 78
LONG NIGHTS, UNSPOKEN PRAYERS ............................................... 78
DAILY LIFE IN IRAQ — THE RHYTHM OF SURVIVAL ....................... 79
CARRYING THE WEIGHT AS A LEADER ............................................ 79
MOMENTS THAT STAY WITH YOU FOREVER ................................... 80
THE GUILT THAT FOLLOWS A SOLDIER HOME ................................ 80
WAR LEFT A MARK — BUT IT DIDN'T END MY STORY .................. 80

## RETURNING HOME — REBUILDING, REFLECTING, AND FINDING MEANING ..................................................... 81

THE AIRPORT REUNION — JOY MIXED WITH SOMETHING ELSE ..... 81
TRYING TO FIT BACK INTO A LIFE THAT KEPT MOVING WITHOUT YOU ...................................................................................................... 82
THE SILENT WEIGHT THAT FOLLOWED ME HOME ......................... 82
TRYING TO BE A HUSBAND AND FATHER AGAIN ............................ 82
THE CONVERSATIONS THAT CHANGED EVERYTHING ..................... 83
SEEING LIFE DIFFERENTLY AFTER WAR .......................................... 83
REBUILDING MYSELF — ONE LAYER AT A TIME ............................ 84
A NEW BEGINNING — FINDING MEANING AGAIN ......................... 84

## WALKING TOWARD FAITH — HOW GETTING TO KNOW JEHOVAH GAVE MY LIFE NEW MEANING .... 85

THE QUIET REALIZATION: THE SPACE WAR LEFT BEHIND .............. 85
TASKA'S EXAMPLE — LOVE, FAITH, AND STABILITY ..................... 86

THE EARLY CONVERSATIONS — QUESTIONS THAT WOULDN'T LEAVE ........................................................................................ 86
THE BIBLE BEGINS TO SPEAK TO ME ............................................. 86
LEARNING JEHOVAH'S NAME — A TURNING POINT ....................... 87
STUDYING THE BIBLE — PIECE BY PIECE, STEP BY STEP ................ 87
FAITH BEGINS TO CHANGE MY MARRIAGE AND FATHERHOOD ...... 88
LETTING GO OF GUILT — A SPIRITUAL HEALING .......................... 88
A SPIRITUAL ANCHOR IN A CHAOTIC WORLD ................................ 88
JEHOVAH GAVE MY LIFE NEW MEANING ...................................... 89

**EPILOGUE- A LETTER TO MY FAMILY ......................... 91**

THIS LETTER—THESE FINAL WORDS—ARE FOR YOU. ..................... 91

**ABOUT THE AUTHOR ............................................................ 95**

# ACKNOWLEDGEMENTS

I would like to express my deepest gratitude to the people who have shaped my journey, lifted my spirits, and helped me become the man I am today.

To my wife, Taska —
Thank you for your unwavering love, your sacrifice, and the strength you carried in silence.
You stood in the gap when I was far from home, held our family together when life was heavy, and believed in me long before I believed in myself.
You are the heart of every chapter in this book.

To my children —
Your smiles, your laughter, and your unconditional love gave me purpose in the darkest moments.
You were the reason I preserved the reason I stayed focused, and the reason I came home with a deeper appreciation for life.

To my family, both near and far —
Thank you for your support, your encouragement, and your prayers throughout my journey.
Your love has always been a source of strength.

To my fellow soldiers —
those who served beside me,
those who stood guard during long nights,
those who didn't make it home,
and those whose battles continued long after the deployments ended —
This story honors your courage, your sacrifice, and your legacy.
To the mentors, leaders, and friends who shaped my career, challenged me to grow, and believed in my potential — thank you for guiding me both as a soldier and as a man.

And above all, to Jehovah —
Thank You for protecting me, guiding me, and carrying me through every season of my life.
Thank You for giving me a second chance at peace, purpose, and home.

This book would not exist without each of you.

# PREFACE

There are memories a soldier carries that have no physical weight, but they bend the soul. For years after I came home from Iraq, I carried the weight of the sounds, the faces, and the agonizing guilt of having survived when others did not. The soldier's mask I wore in the field—steady, calm, and unbreakable—became the mask I wore at home.

This book began not as a public story, but as a form of therapy. It became a necessary process of confronting the silent weight that had followed me home, a way to unpack the emotions that wouldn't fit in any deployment bag. Writing has been my path to rebuilding the parts of myself that were broken during the war.

## The Dual Journey

The story you are about to read is not linear. It follows two parallel journeys:

First, it tracks the professional journey: from a young Jamaican-American kid seeking direction in the Army barracks of Cold War Germany, through the ranks as a medic and NCO, and finally into the highly technical role of a Warrant Officer. These chapters detail the discipline and responsibility required to lead in the military, and the emotional toll of having to point a weapon at a child or hear the final breaths of comrades over the radio.

Second, it charts the spiritual journey: the slow, quiet realization that technical expertise and military strength could not fill the emptiness left by combat. This is the story of how my wife, Taska—my unwavering anchor—helped guide me toward a relationship with Jehovah, who provided the peace and purpose I had searched for across three continents.

**A Note to the Reader**

To help you navigate the military chapters, please note that although I use specific job titles and acronyms—such as 91B, 14J C4I, and 140A—I have always focused on their real-world functions. My job was not just to know systems, but to ensure the Army's eyes and ears saw the same battlefield picture, preventing deadly confusion.

Ultimately, this book is a tribute to the enduring strength of military spouses. Taska's sacrifice, her quiet resilience, and her occasional, honest plea for me to "just get out" are central to this story. You will see her burden and her grace. Without her belief and her faith, there would be no story of coming Home for Good.

# PROLOGUE

Before the Uniform, Before the War, Before the Faith — There Was a Journey I Didn't Yet Understand
There are moments in life that you don't realize are shaping you.
Moments that feel ordinary while they are happening—
a childhood memory,
a father's voice on the phone,
a quiet car ride with someone who will change your life,
a deployment order printed on plain white paper.
Only later do you look back and see the pattern.
The purpose.
The meaning.
For most of my early life, I didn't see it.
I was just a young Jamaican-American kid trying to find his place in the world.
Trying to understand manhood.
Trying to understand responsibility.
Trying to understand himself.
I stumbled through choices.
I learned lessons the hard way.
I walked roads that felt random at the time.
But none of them were random.
Every step, every challenge, every victory, every hardship—
They were all preparing me for the man I would become.
Long before I knew the name Jehovah,

He was guiding my steps.
He guided me through the barracks in Germany,
through loneliness in overseas duty stations,
through moments of discipline and moments of doubt.
He guided me to the woman who would change my entire life with one look in a sun visor mirror.
He guided me through the chaos of fatherhood,
through lessons that humbled me,
through promotions I didn't always feel ready for.
He guided me through Korea,
through the pressure of recruitment duty,
through the challenges of becoming a Warrant Officer,
through Kuwait,
through the morning of 9/11,
through Iraq,
through loss,
through silence,
through survival.
This book isn't a story about war.
It isn't about medals or rank.
It isn't even about the Army, though the Army is part of the path.
It is a story about love—
the love that sustained me, tested me, and rebuilt me.
It is a story about sacrifice—
not just mine, but the sacrifice my wife and children carried without recognition.
It is a story about faith—
a faith I didn't grow up with,
a faith I didn't always understand,
but a faith that waited patiently for me.
It is a story of becoming—not in one moment,
but across a lifetime.

A story of learning what truly matters.
A story of losing myself and finding myself again.
A story of being broken down and rebuilt by something greater than myself.
Most of all-
This is the story of how a soldier
found his true home—
not in a barracks,
not in a deployment ending,
not even in returning from war—
but in the love of his family
and the peace that comes only from Jehovah.
This is the story of how I became,
after everything I endured and survived:
Home for Good.

# EARLY LIFE – JAMAICAN ROOTS & A NEW BEGINNING

I was the only one of my siblings born in the United States, but everything about my upbringing was undeniably Jamaican. From the way my parents spoke, to the food we ate, to the expectations that ran through our home—my childhood rhythm was the rhythm of Jamaica. I didn't realize it at the time, but the way I grew up became the foundation for every chapter of my life that followed. It shaped my discipline, resilience, sense of responsibility, love for family, and ultimately, my faith.

## Growing Up Jamaican in America

Being Jamaican in an American neighborhood meant living in two worlds at once. Inside the house, it was patois, reggae on the weekends, oxtails simmering in the kitchen, and discipline that didn't require long speeches—just one look from your mother could change your whole day. Outside the house, it was American everything—school friends, neighborhood kids, football games, and trying to fit in without losing the identity that made me who I was. Some days, I felt like I belonged everywhere. On other days, I didn't quite fit in anywhere. That tension taught me early how to adapt—a skill that would one day carry me across oceans.

### The Band Room — The Moment I Realized I Was Different

I recall one afternoon in the school band room, talking to my mother on the phone. I wasn't thinking about who was around or how I sounded — I was just being myself. When the conversation ended, I said, "Okay, Mommie." And just like that, my Jamaican accent came out as strong as ever. The whole room went quiet for a second. Some of the other band members looked at me with a mix of surprise and curiosity — not judgment, just the realization that I wasn't quite like them. That was one of the first times I truly felt the difference between my home culture and my school culture. It wasn't a bad feeling, just different. A small moment, but one that stayed with me. It taught me something simple yet powerful: no matter where I went, my roots would always follow me.

### Sunday Breakfasts & School-Day Porridge — The Flavors of My Childhood

I still remember the smell of Sunday morning breakfast, like it's happening right now. Ackee and saltfish sizzling in the pan, green bananas steaming, fried dumplings turning golden brown—the kind only a Jamaican mother can perfect. Those mornings were calm and comforting, love-filled without any words spoken. The day might be uncertain, but breakfast wasn't. On school days, the rhythm was different but just as meaningful. Before we left for school, my mother always made sure we had something warm in our stomachs: cornmeal porridge or rice porridge—thick, creamy, sweet, and seasoned with nutmeg in a way that lingers for your entire life. Even years later, in Germany, Korea, Kuwait, or Iraq, those flavors were a memory that brought me home.

## The Strength & Heart of My Parents

My father worked as the Supervisor of the Food Production Department at Okeelanta Sugar Corporation. That job placed him right in the heart of a community filled with Jamaican migrant workers trying to make a life far from home. They worked long hours, lived in humble conditions, and sent every extra dollar back to Jamaica to support their families. My parents saw them. They recognized their struggle. And they taught us to see people—not for their circumstances—but for their humanity. My father quietly made sure they had food. My mother stretched whatever we had to feed "one more person" when someone knocked on the door unexpectedly. In our home, being kind wasn't a choice—it was a lifestyle. Looking back, I realize that my parents were teaching us the very qualities Jehovah values most—long before I ever learned His name.

### The "Barrel" Tradition — Giving Back to Jamaica

Every year, my parents traveled back to Jamaica, and they never went empty-handed. They brought with them the thing every Jamaican immigrant family knows well: the barrel. A large shipping barrel packed tightly with: clothes, shoes, toiletries, rice, flour, canned goods, school supplies, and little treats for the children. They would take one of us children along each time— not for vacation, but to teach us who we were and where we came from. Those trips taught me the importance of gratitude, kindness, humility, and taking care of your people. Those barrels weren't just supplies. They were love, wrapped in sacrifice.

### Excellence Was the Standard — No C's Allowed

School was serious business in our house. Good grades weren't celebrated—they were expected. I still remember my mother saying one day, with that unmistakable Jamaican tone: "If you bring a C in this house, I'll put you out with the trash." I didn't know exactly what she meant... but I knew I didn't want to find out. The message was clear: Do your best. Work hard. Don't settle. Excellence is the bare minimum. Those expectations shaped me long before I joined the Army. Later in my military career, when boards and promotions came around, I realized I wasn't just trying to meet Army standards—I was trying to meet my parents' standards.

## College — Good Grades, but No Real Purpose

Even in college, I carried that discipline with me. My grades were good, not because I loved what I was studying, but because doing anything less felt wrong. But despite the success, something inside me felt empty. I couldn't shake the feeling that I was working toward something I didn't truly want. It wasn't the work. It wasn't the environment. It wasn't the people. It was the purpose. I kept asking myself: "Why am I doing this? Where is this leading?" And each time, the answer felt the same: "This isn't it." I wasn't lost—but I wasn't found either. I needed direction. I needed structure. I needed a path where discipline meant something, where hard work built a future. The military, with its absolute demand for excellence and structure, felt like the only place that could truly match the high standards my Jamaican parents had drilled into me since childhood. I needed a test, and I knew the Army would deliver it. Something had to change.

## The Day That Changed Everything

One ordinary afternoon, I walked into an Army recruiting office. Not because I had a perfect plan, but because I knew I needed something different—something meaningful. I needed to build a life that was truly mine. I recall the silence of the office, which contrasted with the noise in my head. I sat down, feeling the pull of commitment, yet wondering if this was a matter of desperation or destiny. The recruiter looked at me and asked, "Ever think about being a Combat Medic?" Something inside me clicked—a quiet yes, I couldn't explain. We talked for an hour about the commitment: the years, the discipline, the travel. When the recruiter pushed the enlistment forms toward me, I hesitated, feeling the full weight of the ink I was about to spill. This wasn't signing up for a class; it was signing away my former life. I signed the papers. Just like that. The first call I made was to my parents. My father, predictably, asked me if I'd lost my mind. My mother, however, was quieter, her concern layered with a knowledge that I was finally taking a step toward defining myself.

**Leaving Home for the First Time**

When I left for basic training, I stepped into adulthood in a way college never could have offered. I hugged my family, packed my bags, and walked toward the unknown. I didn't know then that this decision would lead me to: Germany, Louisiana, Fatherhood, Korea, Recruiting duty, Warrant Officer school, Kuwait, 9/11, Iraq, and eventually... to Jehovah. I didn't know about the challenges, the blessings, the heartbreak, or the victories that lay ahead. But I carried with me everything my Jamaican parents had instilled in me—their strength, their discipline, their generosity, their faith, and their love. This was the beginning of my journey, and though I stepped into the unknown, I knew I carried the strength of Jamaica's expectations—the deep-seated belief that I was prepared to meet whatever test lay ahead.

# GERMANY – MY FIRST DUTY STATION, MY FIRST REAL TEST

Leaving home for the first time was a shock to the system, but nothing prepared me for what it felt like to land in Germany—my first duty station, my first time out of the country, my first taste of independence, and my first real test as a young soldier. I was barely out of my teens, still trying to understand who I was, when the Army dropped me into a world completely different from anything I'd ever known. Cold air, unfamiliar faces, a different language, strange foods, and a culture that felt both rigid and fascinating. It was the kind of place that could chew you up if you weren't grounded. Germany didn't care who you thought you were. It forced you to grow up, or it left you behind.

**A Young Soldier in a New World**

I remember stepping off the plane and feeling that cold European air hit my face. It wasn't just the weather—it was the realization: "You're not home anymore. You're on your own." There was no mother waking me up, no father watching my choices, no Jamaican comfort food in the kitchen, no familiar accents around me. Just a group of young soldiers who were just as nervous and unprepared as I was. We didn't know it yet, but we were about to become each other's family.

### Culture Shock in Every Direction

Germany was beautiful—but it could be overwhelming. The language was a challenge. Ordering food, reading street signs, and simply asking for directions became daily puzzles. The cold cut through our uniforms. Winter felt like a force of its own—alive, biting, unyielding. The food was unfamiliar—bratwurst, schnitzel, potatoes prepared every way imaginable. Nothing like the ackee, dumplings, and porridge of my childhood. Homesickness hit me harder than I expected.

### The Bond of Soldiers Away From Home

What made Germany bearable was the presence of the soldiers around me. We were all far from home—from different cities, states, and cultures—but united by the fact that we were young and trying to figure out life together. We ate together, trained together, laughed together, and leaned on each other through the tough days. Those early friendships prepared me for the deeper bonds I'd form later in my career.

### Checkpoint Charlie — The Unforgettable Crossing

My first exposure to the reality of the Cold War was immediate and personal. My duty station wasn't just Germany; it was Berlin, Germany, a city divided by a wall and saturated with tension. To reach my unit, I had to cross through Checkpoint Charlie, the most famous crossing point between West Berlin and East Berlin. The journey was a crash course in Cold War unease. I remember the German soldiers—stern, official, and heavily armed—coming onto the train to inspect our IDs. The silence in the train car was heavy as they moved past us. It wasn't a standard customs check; it was an intersection of two very real, very hostile worlds. While I was there, I called home to check in. My father picked up and said something I will never forget: "Son, I love you." You didn't hear those words often in a Jamaican home—especially from your father. Hearing them while standing in a place defined by the bleak, imposing concrete barrier, the cold, unsmiling gaze of the border guards, and the history of conflict hit me in a way I didn't expect. It grounded me. It reminded me of who I was, even while standing in a world so different from home. Those words carried me through countless challenges that followed.

**Training Under Russian Watch**

The tension wasn't confined to the checkpoints; it followed us into the field. During training exercises, we constantly had a silent, unwelcome audience. We would see Russian soldiers positioned nearby, openly taking pictures of our formations, our gear, and our movements. At first, it was genuinely scary. It was a clear reminder that we were literally training on the front lines, and every maneuver was being documented by the "enemy." But, after a while, you just got used to it. The abnormal became routine. We knew the drill—you could practically time their observations: the rhythmic rise and fall of their surveillance helicopters became just another predictable part of the landscape. This constant observation reinforced the idea that every decision, every movement, carried significance. It made the training feel more real and demanding.

**Learning Responsibility the Hard Way**

Germany forced me to grow up—fast. Nobody reminded me to pay bills. Nobody cooked my meals. Nobody cleaned my room. Nobody told me how to budget, shop, wash clothes, or take care of myself. The Army gave you a roof, but you had to build your life under it. I learned how to: manage money, survive harsh weather, navigate foreign cities, handle loneliness, and make grown-man decisions. I didn't realize it at the time, but Germany was preparing me for everything that would come next.

## Choosing to Become a 91B Combat Medic

Before joining the Army, I was a pre-med student in college. I thought I had it figured out: go to medical school, become a doctor, and build the life everyone would be proud of. So, when the recruiter asked me, "Ever think about becoming a Combat Medic?" I said yes without hesitation. It seemed like the perfect path—medical training, experience, discipline, money for school later. In my mind, the plan was to become a Combat Medic, save money, gain experience, attend medical school, and become a doctor. But the Army had lessons that no classroom could teach. Once I started treating soldiers—seeing fear, pain, injuries, and real emergencies—I realized something profound: The reason I wanted to become a doctor was not the reason I needed to become one. College taught me science. The Army taught me humanity. Being a 91B Combat Medic showed me: healing is not glamorous, compassion matters more than credentials, presence is more important than prestige, service isn't about a title, and you don't need a white coat to change a life. Somewhere along the way, I understood the truth: I wasn't chasing medicine for people—I was chasing the idea of success. Germany humbled me. Serving as a medic shaped me. Those lessons stayed with me throughout every stage of my military journey.

## The Training Accident — The Day I Learned What "Helpless" Really Means

There was a training accident in Germany that I will never forget. I was the medic on site when a young soldier was hit by a mortar round. It wasn't supposed to happen—training is supposed to be controlled, supervised, and predictable. But danger doesn't follow rules. The moment it happened, I heard soldiers shouting: "DOC! DOC! DOC!" In the field, you're not just a medic. You're the closest thing to a doctor those soldiers have. But when I reached him, I knew immediately there was nothing I could do. No bandage, no pressure point, no medical technique could fix what had happened. Standing over a wounded soldier, hands ready but powerless, is a feeling that never leaves you. Helplessness. For a medic, that is the worst feeling of all. It taught me that life in the Army—even in training—can change in a second.

## The Pressure of Being "DOC"

Being a Combat Medic wasn't just about medical skill. It was a dual responsibility. You had to be: as tough as the infantry and medically skilled enough to save them. I marched with them, trained with them, and slept in the field with them. My rucksack was heavier because it carried medical gear on top of everything else. On road marches, I had to stay in formation and also watch every soldier: Are their feet blistering? Are they overheating? Are they hydrated? Are they about to go down? Their lives were in my hands even before a bullet or explosion was ever involved. This responsibility changes you forever.

## The Friendships That Stayed With Me

The soldiers I served with in Germany became more than teammates—they became brothers. We went everywhere together in Berlin, not just for safety reasons, but because we genuinely wanted to. The majority of my friends worked in the aid station with me, and we were all called "Doc" as well. Sharing that title meant we carried a double burden: the pressure of being a young soldier, and the terrifying responsibility of holding another man's life in our hands. We were the ones who trusted each other, leaned on each other, and believed in each other's abilities even when we doubted ourselves. Even after thirty plus years, we are still in contact. The brotherhood created in hardship lasts a lifetime.

## Germany — The Foundation for Everything to Come

By the time I left Germany, I was no longer the unsure young man who stepped off the plane. The Cold War had taught me the importance of urgency, the training accident had taught me the futility of helplessness, and the uniform had taught me the value of humility. I left carrying the unspoken weight of every soldier who would ever call me "Doc."

# MEETING TASKA — THE MOMENT EVERYTHING CHANGED

If Germany shaped me into a man, Louisiana was where my life truly began. Fort Polk wasn't a glamorous duty station. It wasn't the kind of place soldiers dreamed of. But looking back now, I realize it was exactly where I needed to be—because that was where Jehovah placed my wife in my path. Sometimes the biggest chapters of our lives begin in the most ordinary moments.

## Fort Polk — A New Challenge and a New Awareness

Arriving at Fort Polk as a Combat Medic was a new kind of challenge. I thought that, coming from Germany, I had seen it all. But Fort Polk had its own way of testing you emotionally, mentally, and physically. One blessing was that a friend from my Germany unit ended up stationed there too. Seeing a familiar face in a new environment gave me a sense of stability when I needed it most. Fort Polk was a mechanized unit, and at first, I thought that meant less road marching. I was wrong. The work was still intense, the field time long, and the demands heavy. Being assigned to the aid station made things interesting. The oppressive humidity and the constant hum of cicadas were the dominant sensory markers of Fort Polk. We got along well, regardless of our race or background. Black, white, Hispanic, Asian—everyone had a role, and everyone depended on each other. But life has a way of teaching you things if you pay attention. When the work slowed down and people relaxed, I noticed that the aid station would naturally split into groups. Not intentionally. Not out of prejudice. Just naturally. Black soldiers talking together. White soldiers talking together. Hispanic soldiers talking together. It wasn't division—it was simply familiarity. Until Fort Polk, I hadn't realized how people instinctively gravitated toward what they knew. Where they felt understood. Where they felt comfortable. But when a soldier needed help, when someone was hurt, when the mission kicked in—all those invisible lines disappeared. In those moments, we were one team. One unit. Family. This awareness of people's need for genuine connection was already humbling me, but I hadn't yet realized that the deepest connection—the kind that transcended any line—was waiting just around the corner.

### A New Chapter Opening Without Me Realizing It

Fort Polk challenged, matured, and prepared me in ways that Germany never did. But even with all that growth, I had no idea my life was about to change in a completely different way—not because of the Army, but because of a woman Jehovah placed right in my path at exactly the right time.

It was also during this period that I gained a new brother, a friend who was there from the moment I met Taska and who, decades later, became the inspiration for me to write this book. We did pretty much everything together, from double dating with our wives—our current partners of over 30 years—to linking up recently. These types of friendships you really never take lightly.

### The Car Ride That Changed My Life

The day I met Taska wasn't planned—it wasn't even something I saw coming. I had wrecked my car, and a friend offered to take me around to look for another one. He asked casually: "Hey, you mind if someone rides with us?" I didn't think anything of it. "Sure," I said. And then she got into the car. She sat in the front passenger seat, and I took the seat right behind her. At first, it was just a normal ride—looking at used cars, talking casually, passing the time. But then something happened. As we were driving, she pulled down the sun visor mirror and looked at me through it—not intentionally, not flirtatiously, just naturally. And in that small, simple moment, something quiet and certain hit me deep inside. I didn't know her story. I didn't know her heart. But something inside whispered: "Pay attention. This one is different." It's funny how life works. You can live years without clarity, and then understanding arrives in a heartbeat.

## A Connection That Became Part of My Every Day

After that car ride, we found ourselves talking more. Then every day. Then seeing each other every day. No awkward pauses. No, trying to impress each other. No games. Just natural conversation. Ease. Comfort. Like I had known her long before that day. There was a warmth about her—a quiet confidence, a kindness in the way she listened, a gentleness balanced with strength. Before long, I realized I didn't want a day to go by without hearing her voice.

## The Two-Day Silence — When I Realized I Couldn't Live Without Her

Not long after we met, we had our first real argument. I can't even remember what it was about now, but at the time it seemed big enough that we didn't talk for two days. Two days. It doesn't sound like much, but when someone becomes a part of your everyday life, two days feels like forever. I kept going back and forth with myself: "You need to call that girl." "No, I'm not calling her." "You need to call her." "No, she should call **me.**" Back and forth. Back and forth. A battle between pride and truth. But underneath that stubbornness was a realization I couldn't ignore: I didn't want to imagine my life without her. Those two days told me more about my feelings for her than any conversation ever could have. That silence showed me she wasn't just someone I cared about—she was someone I couldn't lose.

### Two Weeks Later — A Realization I Couldn't Ignore

Most people say love takes time. Months, years, slow building. But two weeks after that first car ride, I knew. Not in a flashy, fairy-tale way—but in a deep, steady, unshakeable way. Every moment with her made sense. Every day felt better than the last. Every conversation told me something: "You can't let this one go." That certainty humbled me. It scared me. But it also made everything clear.

### The Proposal Test — A Choice That Defined Everything

Taska had only come to Louisiana for the summer after graduating high school to be a nanny for a friend. Her stay was temporary, and the date for her return to Alabama was fast approaching. I didn't want her to leave without knowing how I felt, but more importantly, I didn't want her to leave without asking her to stay forever. One day, I told her, "If you're here when I get back, we'll go pick out your ring. And if you're not... that'll be my answer." It wasn't a threat. It wasn't a game. It was truth. If she stayed, it meant something real. And if she didn't, I would know I was wrong. When I pulled up to the house, she wasn't there. For a moment, I felt a small ache in my chest. Maybe I moved too fast. Maybe I had it wrong. But just as I was about to pull away, a car pulled into the driveway. It was her. She didn't hesitate. She didn't walk slowly. She didn't think twice. She jumped right into my car. And in that moment—before rings, vows, or promises—she chose me. No uncertainty. No confusion. She chose me.

## How I Knew She Was the One

People often ask how you know when you've found the right person. For me, it was simple: She carried the same qualities I had grown up admiring in my parents—the compassion, the generosity, the instinct to care for others, the quiet strength wrapped in gentleness, and the ability to love without show or pretense. She felt like home before she ever became my home. Meeting Taska didn't just change my life—it redirected it. Everything that came after was shaped by her presence, her love, her patience, and her belief in me. And as soon as we took that step together, life began unfolding in ways neither of us could have imagined.

# STARTING OUR LIFE TOGETHER — YOUNG, MARRIED, AND LEARNING AS WE WENT

Starting a life together is never simple. Starting a life together as young newlyweds in the military, far from family, without much money, and with responsibilities already growing—that is something else entirely. But we were young, in love, and believed that if we had each other, we could figure out the rest. And in many ways, we did.

## Our First Home — Humble, Colorful, and Ours

Our first home wasn't a house at all—it was a two-bedroom trailer, small and simple, but full of the excitement that comes when two young people start their life together. Money was tight, but one thing we took pride in was our furniture. Everything was brand new—not hand-me-downs, not borrowed, not pieced together. Brand new. And colorful. Our living room looked like a burst of energy—bright, bold fabric with patterns loud enough to speak before you did. Even the coffee table and end tables had personality, with colors and designs that only young newlyweds with big dreams would confidently pick. Looking back, it makes me smile. We didn't have much money, but what we did have was ours—chosen with hope, bought with pride, and arranged with excitement in that little trailer. It may have been small, but the future we imagined inside it felt big.

### Learning How to Be Married

Every married couple has their learning curve. Ours began the moment we carried our boxes inside. We were learning everything from scratch: how to budget together, how to combine two different upbringings, how to communicate without shutting down, how to argue and come back together, how to split responsibilities, how to support each other through long military days, and how to build routines that worked for "us" instead of "me." Marriage wasn't just love. It was growth. Day by day. Lesson by lesson. Those early days shaped us more than we realized.

### Money Was Tight, but Love Was Strong

People talk about "broke love," but until you've lived it, you don't fully understand it. We weren't poor, but we were far from comfortable. Young soldiers pay didn't stretch very far, and we had to learn how to make every dollar count. Some weeks, we had to choose between eating out and saving for bills. We learned fast that: home-cooked meals go a long way, off-brand groceries taste just as good, laughter helps when the money doesn't, teamwork makes tight seasons bearable, and love doesn't need luxury to survive. Those early struggles built our foundation. They taught us discipline, humility, and how to prioritize what mattered most. I would often think about my father, observing how he managed their household, and tried to model his quiet, steady dedication.

## A Young Soldier, A Young Wife — Two Lives, One Mission

Being married to a soldier is not like being married to anyone else. Your time is not your own. The schedule changes overnight. Field exercises come without warning. Shifts run late. Plans change. Sleep is limited. But Taska handled it with a grace far beyond her age. She never complained when the Army's needs took precedence over plans we had made. She understood before I fully did that service was a commitment we both had to carry. I would sometimes come home extremely stressed after two-week stints in the field—exhausted, dirty, and mentally drained. All she would do, before anything else, was spend a little time with me, grounding me immediately in our home and our life together. She didn't just marry me—she married the lifestyle. And through her patience and strength, she grounded me in ways I didn't even know I needed.

### Learning to Be a Husband

Growing up Jamaican, I saw marriage through a very specific lens—structured roles, discipline, and expectations. But becoming a husband is very different from watching one. I had to learn how to listen, communicate through frustration, apologize, lead with love, support emotionally, not just financially, show patience, love in a way she understood, and be present even when exhausted. Those weren't always easy lessons, but they were necessary ones. Marriage didn't just teach me about her—it taught me about myself.

### Building a Partnership, Not Roles

We came from different backgrounds, different cultures, different ways of seeing the world. Early on, we had to learn how to blend those worlds. There were moments where her expectations clashed with what I was raised to believe, and moments where mine challenged hers. But instead of letting those differences divide us, we learned to talk, to listen, to understand. We weren't trying to win. We were trying to grow. And that mindset became one of the greatest strengths of our marriage.

**Laughing, Learning, and Loving Through the Struggles**

Despite the challenges, there was so much joy in those early years. Inside jokes. Late-night conversations. Simple moments that felt like home. Daily routines that became memories. Small victories that felt huge. Those moments reminded us why we chose each other. We learned to enjoy the quiet. To make regular days meaningful. To celebrate the small things. To build a life filled with love, even when circumstances were tight. Our beginnings weren't glamorous, but they were beautiful. They were ours.

**Preparing for the Next Chapter — Becoming Parents**

Even while we were still learning how to be married, life was quietly preparing us for our next role: mother and father. We didn't know it yet, but everything we had learned so far—patience, communication, sacrifice, and teamwork—was training for what was to come. Becoming parents would stretch us, grow us, and transform us in ways we couldn't imagine. But everything started here—in that colorful, humble two-bedroom trailer, with big dreams, young hearts, and a love still taking shape.

# BECOMING A FATHER —
# THE BEGINNING OF A NEW LEGACY

Nothing prepares you for the moment you find out you're going to be a father. It doesn't matter how old you are, how strong you think you are, or how much life you've already lived—that moment changes everything. For me, it was a shift so deep that I felt it in my chest before I even understood it in my mind. One day, we were a young couple figuring out life in a two-bedroom trailer. Next, I found out we were going to bring a whole new life into the world. And suddenly, the world felt bigger—and I felt smaller. Not in a powerless way, but in a humbled, "this responsibility is bigger than me" way.

## Hearing the Words That Changed My Life

I remember the moment Taska told me she was pregnant with our first child. I felt shock, excitement, fear, joy—all mixed into one moment that I can still feel today. I was about to become responsible for a whole human being. Someone who would call me "Daddy." Someone who would look to me for guidance. Someone whose whole world would be shaped by my choices. It was overwhelming—but also beautiful. For the first time in my life, I wasn't just living for myself. Every decision from that day forward would impact my family. And that realization changed me.

## The Birth of Our First Child: A New Kind of Love in Texas

My unit moved from Fort Polk, Louisiana, to Fort Hood, Texas, where our first daughter, Pa'tricka (Tricka), was born. When Tricka was born, my heart opened in a way I didn't know was possible. Holding her for the first time, I felt a kind of love that had nothing to do with romance, nothing to do with passion, and everything to do with protection, commitment, and purpose. She was so small, so perfect, so innocent—and she depended on us completely. Her tiny fingers wrapped around mine, and something inside of me said, "This is what you were made for." I wasn't just a husband anymore. I was a father. And everything changed.

## A Necessary Evolution: From Medic to Maintainer

The responsibilities of fatherhood—the urgency, the permanence, the financial pressure—crystallized my need for a new direction within the Army. The intense, emotionally charged world of a 91B Combat Medic, while invaluable, felt less compatible with the stability my growing family required. About a year after Tricka was born, we realized we were pregnant again, and I made the impulsive choice to get out of the Army for about a month. The decision was born out of intense financial anxiety and the stress of new parenthood. The civilian world quickly reminded me that the certainty and structure of the Army were what my growing family needed most, particularly when I looked at the instability of the jobs available to me. When I quickly re-enlisted, I had to change my MOS, transitioning from a 91B Medic to a 14J Air Defense C4I System Integrator. This was a profound shift from treating physical trauma and human vulnerability to managing data infrastructure and command systems. It brought with it a compelling realization: As a medic, you learned to read soldiers—to distinguish genuine pain from the malingering soldier who wanted to avoid duty. Humanity, for all its strength, had a capacity for deception. I realized people could fake an illness, but a computer could not. This new technical role—managing the complex, binary logic of C4I (Command, Control, Communications, Computers, and Intelligence) systems—offered a certainty and reliability that, after the emotional demands of being "Doc," was deeply grounding.

## Two More Blessings in Colorado

We were then stationed at Fort Carson, Colorado, with its sharp, thin mountain air and a backdrop dominated by the dramatic peaks of the Rockies. Fort Carson became the birthplace for our second daughter, Tionna, and our son, Joslyn Patrick. Fifteen months after Tricka arrived, we welcomed Tionna—another blessing, another reason to grow, another piece of our legacy. Fifteen months later, our son JP arrived—fifteen months after his sister. After two perfect, precious daughters, there was a quiet, profound moment when I held my son for the first time. He was an answer to a prayer I didn't even know I was praying—a piece of myself, a new link in the chain, a small, trusting face that would someday share my name and carry a piece of my spirit. I looked at him and felt the legacy begin anew, this time with a deep, silent sense of completion. Three babies in such a short time turned our world into a beautiful kind of chaos. It was exhausting. It was nonstop. It was unforgettable. We didn't have time to overthink things. We were too busy living them. I watched Taska become a mother three times in less than three years—and the strength she showed was unlike anything I had ever seen. While I served, trained, and fulfilled my duties, she was at home raising our babies with patience, gentleness, and a courage that still amazes me today. The small size of our military housing amplified the sheer volume of our lives—it felt like a permanent, wonderful pressure cooker.

## The Weight of Responsibility — and the Strength It Built in Me

Becoming a father shifted my priorities overnight. Suddenly, every decision mattered: how I worked, how I trained, how I saved, how I spent, how I planned, and how I carried myself as a man. My children became the measure of my choices. Their future depended on my discipline, career, effort, and presence. There were moments when the weight felt heavy—especially as a young soldier still finding my way. But alongside the weight was something else: purpose. Fatherhood gave me direction in a way nothing else had. It made me more responsible, more focused, more determined. It anchored me.

## Watching Them Grow — The Joy That Made Every Sacrifice Worth It

In those early years, our home was loud, lively, and full of moments that shaped our family forever. I remember: Tricka's baby giggles echoing through the trailer, Tionna's little feet running across the floor, crayon marks on walls, toys everywhere, tiny voices calling, "Daddy!" when I walked in, the way all three wanted to climb into my arms at once, and family nights where love filled every corner of the house. Those memories stayed with me through every deployment, every training mission, every night I slept in a tent half a world away. Fatherhood became my source of strength long before I understood the concept of faith.

## The Foundation of My Future — Built on Love

Before my military career reached its height, before promotions and boards, before deployments, before Warrant Officer school, before Kuwait, before 9/11—I became a father. And that shaped every chapter that followed. My children gave me a reason to push harder, to climb higher, to stay strong when the world grew heavy. They were my motivation, my joy, my grounding, my legacy. And those early years—with three babies only 15 months apart—built the foundation of who I would become as a husband, as a soldier, as a leader, and ultimately... as a man trying to find spiritual truth.

# CLIMBING THE ENLISTED RANKS — BOARDS, BOOKS, AND BREAKTHROUGHS

Growing as an enlisted soldier wasn't just about doing your job. It was about proving yourself—to your leaders, to your peers, to your unit, and most importantly, to yourself. Each stripe, each rank, each evaluation represented more than achievement. It represented discipline, responsibility, and growth. And with a young family depending on me, the stakes were higher than ever. I wasn't climbing the ranks for pride—I was climbing them for purpose.

**A Soldier with Something to Prove**

By the time I reached this stage of my career, I had already learned hard lessons in Germany, survived challenging days at Fort Polk, become a husband, become a father—three times over, and discovered a deeper sense of responsibility. Every promotion opportunity felt like another step toward providing stability for my family. This desire was amplified by the temporary separations we experienced, including deployments to Cuba while stationed at Fort Carson, which brought home the reality of service. Each evaluation mattered. Every decision mattered. Every training moment counted. It wasn't just about being a good soldier—it was about being a dependable man.

## Studying for the Board — My Late-Night Classroom

Anyone who has ever gone before a promotion board knows the stress that comes with it. Uniform perfect. Shoes shining like glass. Brass polished. Hair cut sharp. Voice steady. Bearing professional. But none of that mattered if you didn't know your material. FM manuals, MOS-specific knowledge, history, values, regulations—you had to know it all and then some. And every night, after long days of work and family life, I studied. But I didn't study alone.

## Taska — My Unofficial Drill Sergeant

Night after night, Taska sat with me at our small kitchen table, quizzing me like she was preparing me for the SAT and the NBA Finals at the same time. She would hold the book, look at the questions, and fire them at me without mercy. "What's the maximum effective range of the M16?" "What are the seven Army values?" "Who is the Sergeant Major of the Army?" "Recite the NCO Creed." "What's the first step in treating a casualty?" If I hesitated—even for a second—she asked it again. Sometimes with a smile. Sometimes with a look like, "You better get this right." At first, I didn't realize it, but she was training me with a level of intensity only someone who believed in me could offer. She took my goals personally. My success became our mission. And every night, surrounded by books, flashcards, diapers, bottles, and the sound of three kids sleeping, we prepared for my future.

## The First Board — Nerves and NCOs

Walking into my first NCO Board and E-5 promotion board at Fort Carson, where I served as an E-4/Specialist (SPC), was like stepping into a spotlight with every mistake ready to be exposed. A row of senior NCOs staring at you. Your heart beating through your chest. Your uniform is sharp enough to cut glass. No room for error. You salute. You take your seat. And then the questions come—fast, technical, precise. But because of those nights with Taska, I answered with confidence. And when the board president nodded, when they said I performed well, when I walked out of that room feeling like I had just taken another step toward being the man I wanted to be—I knew we did it. Not just me. We.

## Promotion to Sergeant

The culmination of those efforts came shortly after. I was officially promoted to E-5, Sergeant, and Taska was the one who pinned the new rank onto my uniform. I remember the moment clearly: the weight of the stripes, the honor of the promotion, and the fulfillment of a goal that was truly a team effort. But what I remember most was the look on Taska's face—how proud she was that I had earned that rank and the responsibility that came with it. It wasn't just a military milestone; it was a promise kept to my family. It wasn't luck. It wasn't chance. It was preparation. It was marriage. It was teamwork. It was purpose.

## Leadership — Growing Into the Boots I Wore

As I climbed the ranks and completed mandatory training, such as the Primary Leadership Development Course (PLDC), I learned something important: leadership isn't about stripes. It's about service. Being an NCO meant putting soldiers first, taking care of those who worked under me, doing the hard jobs, making decisions that mattered, setting an example, training others, being a 14J C4I operator and a leader, carrying responsibility, and building trust. The Army taught me the structure of leadership. Marriage and fatherhood taught me the heart of it. Combining both made me a better soldier and a better man.

## The Drive to Keep Climbing

Every achievement pushed me to the next one. Not because I wanted stripes—but because I wanted to provide for my family and honor the sacrifices Taska made to help me succeed. I had come far from the young man who joined the Army searching for purpose. Now I had a purpose—three little children calling me Daddy and a wife who believed in me more than I believed in myself. And that purpose would carry me all the way through my next major assignment—the unaccompanied tour to Korea—into recruiting duty, into deeper responsibilities, and eventually toward a path neither of us could have predicted—the path to becoming a Warrant Officer.

# KOREA — A YEAR AWAY FROM MY FAMILY

Nothing prepares you for being separated from the people you love most. You can be disciplined, trained, strong, and focused—but when that plane lifts off and you look down at the world below, leaving a wife and three babies behind, your heart feels it long before your mind does. Korea was my first major assignment away from my family, and it changed me in ways no training manual ever could.

## Leaving Everything That Mattered

When I boarded that plane to Korea, I left behind: a young wife, three children, all 15 months apart, a home we had built with love and sacrifice, the noise, laughter, and chaos of parenthood, and my daily identity as "Daddy."

The sheer dread of that separation was never clearer than during my final preparations. I remember Taska telling me, "You can just get out." I could see the fear in her eyes—the fear of being lonely and having to raise our three children by herself. It was too much to ask of a young couple: not raising our kids together, even if it was just for a year.

To give her some help, we moved the family back to Alabama, where she could lean on her Aunt Gayle and her stepdad, Ralph.

The Army trains you for missions, but it doesn't train you for the quiet ache of missing home. Alone in the barracks, I would ask myself, *Does this loneliness have a greater purpose, or am I simply sacrificing time I can't get back?* While other soldiers were excited about an overseas tour, I felt torn between duty and family. I carried their faces with me everywhere I went. I had sent a letter home to my mother, knowing she understood sacrifice, but her worried call only amplified my own doubt.

Taska, in truth, was a soldier—not in the military sense, but as all military spouses know, you are some of the strongest people and don't get enough credit.

### Arriving in Korea — Cold, Busy, and Lonely

Korea was alive with energy—bright lights, fast movement, and a pace that never slowed down. But inside, I felt alone. No familiar accents, no small feet running across the floor, no nightly routine with the kids, no warm conversations with Taska after a long day. Just barracks walls, a narrow bed, a desk barely big enough for my elbows, and long days that made nights even longer. I was a soldier by day—but at night, I was a husband and father struggling with distance.

The separation hit hardest in the small things: not being there for the kids' milestones, missing first words, first steps, hearing their voices change through the phone, and getting letters with drawings. Listening to Taska carry the weight alone was agonizing. The ultimate test came when Pa'tricka got seriously sick; the helplessness of being thousands of miles away while Taska navigated the worry and doctor visits alone was agonizing. But as always, Taska held it down. Every time I hung up the phone, the silence afterward felt heavy. And every night before I fell asleep, I prayed for their safety, their health, and for strength for my wife. I would try to imagine her days: the exhausting routine of three babies, the late nights, the sudden illness. I could almost read the letter she never sent—one filled with worry and fatigue—a silent counterpoint to the brave, cheerful voice she always saved for my calls.

## Taska — Carrying the Weight at Home

While I was thousands of miles away wearing the uniform, Taska was at home wearing a different kind of armor. She took care of Pa'tricka, Tionna, and JP, managed the house, handled the bills, attended appointments alone, fixed meals alone, tucked the kids into bed alone, and woke up for late-night cries alone. One night, she admitted during a late-night call, "I'm scared, Patrick. This is just too much sometimes." I remember hearing the faint sound of JP crying in the background, a reminder of the non-stop battle she was fighting just to keep things normal. Her bravery was in her consistency, but her true burden was the loneliness of that strength. She was my stability even when I couldn't be there. She would tell me, "Focus on your job. We're okay." But I knew it wasn't easy. I knew she carried more than she ever admitted. And that gave me even more determination to make something of myself in Korea.

## The Staff Sergeant Board — Pressure and Purpose

One of the most significant moments in Korea was appearing before the Staff Sergeant (E-6) board. Another soldier and I went to the board together. We had trained hard, studied nonstop, and pushed each other with friendly competition. And even thousands of miles away, Taska still prepared me: quizzing me over the phone, encouraging me when I doubted myself, and reminding me why I needed to succeed. When the results came out, and I learned

I was being promoted, I felt a flood of emotions—pride, relief, and gratitude. But more than anything, I wished Taska had been there to see the moment. Because it wasn't just my achievement—it was ours.

## Daily Life in Korea — A Routine of Duty and Discipline

Korea had its challenges, but it also had its rhythms. My assignment was particularly challenging: I was detailed as a liaison at the Aviation Brigade (AVN BDE), the only Air Defense personnel in the unit. I had to learn how the aviation attack helicopters worked within the overall air defense and airspace management structure. This forced me to quickly become a subject matter expert in bridging two different worlds. I filled my days with: PT in the mornings, long workdays managing the crucial C4I air defense systems and consulting on joint airspace management, field exercises, medical training (to maintain that critical skill), staying busy, keeping my mind engaged, writing letters, talking to my kids when I could, calling Taska late at night, and counting down the days until I could go home. I made good friends there—soldiers who understood the loneliness, the pressure, and the responsibility. We leaned on each other because we all carried burdens we didn't speak about openly.

## The Emotional Distance — and the Love That Survived It

Being apart for a year tested our marriage in ways we'd never experienced. There were lonely days, frustrating misunderstandings, missed calls, and moments where we felt the distance too deeply. But there were also reminders of why we worked: handwritten letters, small gifts mailed across the oceans, late-night calls with the kids, shared goals, shared dreams, and the determination to stay strong for one another. The distance didn't break us—it strengthened us. It made us fight harder, communicate deeper, and appreciate each other more.

## Returning Home — A Man Changed by Distance

When my tour ended and I finally stepped off the plane back onto American soil, everything felt different. I wasn't the same man who had left. Korea had matured me. It had humbled me. It had strengthened me. It had reminded me of my priorities. It had deepened my love for my family. And the moment I saw Taska and my children waiting for me, it felt like my heart was being put back where it belonged. That year tested our bonds through distance, and in surviving it, Taska and I proved that our commitment was stronger than any duty station.

# RECRUITING DUTY — PRESSURE, PURPOSE, AND CHANGING LIVES

After returning from my unaccompanied tour in Korea, I expected life to slow down a little. I expected a break—time to reconnect, to rest, to breathe. The Army had other plans, but first, it offered a choice.

## A Warrant Officer Path Deferred

I was reassigned to Fort Polk, Louisiana, and promoted to Staff Sergeant (E-6) during transit. Life finally seemed set to stabilize, but the Army had a surprising intervention. While at Fort Polk, I received a letter encouraging me to submit a Warrant Officer (WO) packet. Honestly, I was skeptical. That path seemed reserved for others, not for a newly promoted NCO from a technical background. I wasn't sure if I was ready, or even qualified. I voiced my hesitation, but a Chief Warrant Officer Four (CW4) pulled me aside and offered some unforgettable advice: "You need to put your packet in. The Army sees something in you." I knew he was right, but my skepticism and uncertainty led me to delay. Because I did not submit my packet in a timely manner, the Army took that opportunity off the table—for now—and sent me directly to Recruiting Duty in Clovis, New Mexico. It was a place I had never imagined living. Flat land. Open skies. Wind that never stopped. Dust storms that could change the color of your car by evening. But Clovis would become a chapter that tested me in ways the field never did—and shaped me in ways I still carry today.

## A World Away from the Army I Knew

Recruiting duty felt like a different universe. No formations. No barracks life. No field exercises. No platoon to lean on. No squad leader watching over you. Just you, a government-issued laptop, a monthly mission, and the weight of representing the United States Army to an entire community. Suddenly, I wasn't "Doc" anymore. I wasn't the medic soldiers ran to. I wasn't in the field. Out here, I was simply a Sergeant. That title, earned through hard work and Taska's help, carried immense weight in the community—it was a tangible symbol of stability and success. I was wearing dress blues in high schools, college fairs, parking lots, front porches, and football stadiums. I was selling a dream some kids desperately needed—and some desperately wanted to escape.

## The Pressure — Numbers, Numbers, Numbers

Recruiting duty came with a pressure that felt different than anything in uniform. Every month had a mission: how many contracts, how many high school grads, how many ship dates, and how many test qualifiers. Your whole career rested on whether you "made mission." Miss it too many times, and your reputation suffers. Hit it consistently—and you thrived. There were days when I felt like a counselor, days when I felt like a mentor, days when I felt like a father figure, and days when I felt like a salesman trying to meet impossible expectations. But what kept me going was not the numbers—it was the people.

## Helping Young People — Finding Purpose in the Hard Days

Clovis may have been a small town, but it was filled with young people who were lost, confused, or simply desperate for a chance to build a better future. The landscape itself—flat, wide, and open—felt like a place where futures could easily get lost in the horizon. Many came from homes with little guidance. Some had no direction. Some were on the wrong path. And every once in a while, you'd meet someone you knew you could help transform. I still remember sitting across from young men who didn't believe in themselves. Who thought their mistakes defined them? Who thought their future was limited to whatever they saw around them? I saw myself in some of them—young, searching, hungry for purpose. And I found myself saying things that came from deep inside: "You're better than the streets." "You're smarter than your environment." "The Army won't fix everything—but it will give you a chance to fix yourself." For the first time, I realized I wasn't just enlisting soldiers—I was impacting lives.

### Parents Who Thanked Me

Some moments will stay with me forever. There were parents who cried when their sons and daughters signed. Not because they were losing them—but because they were gaining hope. Mothers hugged me with tears running down their cheeks, saying, "Thank you for believing in my child when they didn't believe in themselves." Fathers shook my hand and whispered, "You saved my boy." Parents who had been exhausted by their child's behavior saw the Army as a bridge to responsibility, structure, and pride. In those moments, the pressure, the long hours, the stress, the quotas—none of it mattered. I realized recruiting was more than numbers. It was a purpose.

### The Hardest Part — When They Didn't Make It

But it wasn't all success stories. Some young people I worked hard to help didn't pass the ASVAB. Some failed medical screenings. Some didn't qualify for moral reasons. Those were the days that weighed heavily on me. You want so badly to give them a chance to change their life, but you can't bend the rules. Those moments taught me the hardest truth of recruiting: You can't save everyone. But you can give everyone the respect and honesty they deserve.

### Long Hours, Long Drives, Long Days

Recruiting duty had a rhythm: high school visits, classroom talks, home visits, parent meetings, late nights in the office, early morning PT sessions with recruits, traveling across rural towns, and paperwork that never seemed to end. Days that started before sunrise and ended long after sunset. It wasn't glamorous. It wasn't easy. But it was meaningful. It required discipline, patience, and a level of professionalism that built skills I would use for the rest of my career.

### Taska's Strength During This Chapter

While I was out late meeting monthly missions, driving across counties, and guiding young people, Taska was at home holding everything together. She was raising Tricka, Tionna, and JP, managing the house, helping with homework, cooking meals, balancing the budget, and grounding me when the month got stressful. Her support even extended to the mission: she would often speak with the future military spouses of my recruits, offering them advice, sharing her experiences, and easing their fears. She was a counselor and a partner in the most unexpected way. Despite her unwavering support, I knew the strain of those long, unpredictable hours weighed heavily on her; sometimes, late at night, she would admit she was exhausted and missed the stability we used to share. But she stood firm, steady, and supportive through every late night and every early morning. She never made me feel guilty for the hours I had to put in. She understood the mission. She understood me. She was my steady place in the chaos of monthly missions.

## Moments That Stay With Me

There were successes that made everything worthwhile: watching a recruit transform from unsure to confident, seeing them complete basic training, receiving letters saying, "Thank you for pushing me," and parents hugging me with pride in their eyes. Despite the short time I spent on recruiting duty—less than a year—I was able to receive the Gold Badge, a tangible recognition of superior achievement and mission success. The pride I felt in that moment was immense, knowing I had helped change so many lives and met such a demanding standard. Those are the moments that made recruiting duty unforgettable. Those are the moments that reminded me why the Army chose me for that assignment.

## Recruiting Duty Prepared Me for Something Bigger

At the time, I didn't realize it—but recruiting duty was shaping me for the next chapter of my life. It refined my: communication, public speaking, counseling, leadership, confidence, discipline, professionalism, patience, and understanding of people. Skills that would later become essential in my journey toward becoming a Warrant Officer. Clovis challenged me, stretched me, and matured me in ways I can't fully describe. It strengthened my marriage, made me a better father, and deepened the purpose behind my military career.

# THE WARRANT OFFICER JOURNEY — STEPPING INTO A NEW LEVEL OF LEADERSHIP

Every soldier has a moment when they stand at a crossroads and feel the pull of something bigger. For me, that moment came after recruiting duty—after long months of pressure, life-changing conversations with young people, and the personal growth that only comes from helping others find their purpose. But now, it was time for me to step into something greater. It was time for me to step into leadership at a whole new level. It was time for the Warrant Officer journey.

## The Decision — A Quiet Voice Pushing Me Forward

Becoming a Warrant Officer wasn't something I grew up dreaming about. It wasn't even something I fully understood early in my career. But as I grew professionally—as a medic, as an NCO, as a recruiter—something inside me began to shift. I felt a pull toward deeper responsibility. Toward mastering my craft. Toward leveraging my experience to help shape systems, teams, and missions bigger than myself. The Army needed more leaders in technical roles. More experts. More integrators. More problem-solvers. And little by little, I started seeing myself not just as a soldier, not just as "Doc," not just as an NCO—but as someone ready to move into that next level.

### Taska — The Voice That Pushed Me Forward

When I told Taska I was thinking about applying for Warrant Officer school, her answer was immediate: "You were made for this." Not "maybe." Not "you'll try." Not "let's see." "You were made for this." She believed in me in ways I sometimes struggled to believe in myself. She saw my leadership. She saw my dedication. She recognized my ability to solve problems, help others, and remain calm under pressure. She saw the Warrant Officer before I ever wore the rank. And her belief became the fuel that carried me through the journey.

### Applying — The Weight of the Unknown

Applying to become a Warrant Officer was a process filled with paperwork, recommendations, evaluations, physicals, background checks, nerves, and fear of the unknown. Would I be accepted? Was I ready? Was I good enough? Did I truly belong at that level? These questions stayed with me. But every time I doubted, Taska reminded me: "The Army needs good NCOs, yes... but it also needs good Warrant Officers."

### Warrant Officer School — Humbling and Transformational

WOCS (Warrant Officer Candidate School) wasn't just about academics. It was about discipline, bearing, confidence, humility, teamwork, leadership, and self-awareness. It pushed me harder than many things I had done in uniform. But there was one moment that stayed with me. A TAC officer looked me dead in the face and said, "Plummer, the Army still needs good NCOs. Are you sure this is where you belong?" For a split second, that doubt hit me hard. Was I leaving the part of the Army where I had truly excelled? Was I stepping into something beyond my abilities? Was I trying to be something I wasn't meant to be? But then, in the quiet of that thought, another voice inside me said: "Yes. You are ready." This journey wasn't about ego. It wasn't about rank. It was about growth. I reminded myself of the soldiers I'd helped, the responsibilities I had carried, the missions I had supported, the trust others placed in me. And I remembered Taska's words: "You were made for this."

## Becoming a 140A — A New World of Responsibility

When I graduated and earned the Warrant Officer rank, I stepped into one of the most complex and technical fields in the Army: 140A — Air and Missile Defense Systems Integrator. This was the logical progression from my E-6 role as a 14J C4I System Integrator (Command, Control, Communications, Computers, and Intelligence); I was transitioning from an operator and maintainer to the chief technical advisor and architect. This job wasn't simple. It required a deep understanding of tactical data links, system integration, joint command and control, air defense coordination, early warning systems, C2 architecture, digital networks, multi-service operations, missile and airspace management, technical leadership, security, and precision that could impact entire missions and lives. In simple terms, my job was to ensure the Army's eyes and ears—the radars and missile systems—were all seeing the same battlefield picture simultaneously, thereby preventing deadly confusion. I became responsible for ensuring that massive systems communicated with each other—that radars, missile systems, aircraft, and command centers shared the same picture simultaneously. In other words, I had become the integrator of the battlefield's eyes and ears.

## Leadership at a Higher Level

As a Warrant Officer, I wasn't just doing my job anymore. I was shaping missions, training soldiers, advising commanders, preparing units, and standing between confusion and clarity in combat operations. I served as a technical expert, senior advisor, problem-solver, teacher, communicator, integrator, and trusted voice in the room. It was leadership on a whole different level. My decisions mattered. My recommendations mattered. My ability to connect systems and correct problems could save lives. This was no longer the young medic running to wounded soldiers in Germany. This was a senior technical leader trusted with complex strategic systems at the heart of the Army's Air and Missile Defense mission.

## The Humility Behind the Rank

But even with the rank, the responsibility, and the trust—I never forgot where I started. I never forgot: being that young soldier stepping off the plane in Germany, the training accident that taught me the weight of helplessness; becoming a husband in a two-bedroom trailer, where fatherhood reshaped my purpose; recruiting duty, teaching me that people matter more than numbers; and Korea, teaching me discipline and sacrifice. My journey was built one step at a time, one challenge at a time, one blessing at a time. The Warrant Officer rank didn't change who I was—it revealed who I had been becoming all along.

## Taska's Role — The Quiet Strength Behind the Achievement

As I stepped into this new chapter, the truth was clear: I didn't get here alone. Taska had sacrificed her schooling, her plans, her dreams, her time, her sleep, her peace—to support mine. She held our family down through every field exercise, every late night, every deployment, every training cycle. She pushed me when I doubted myself. She corrected me when I was too hard on myself. She motivated me when the path felt long. She loved me through every step of the journey. Behind every achievement I gained was her steady hand. Behind every promotion was her quiet encouragement. Behind every success was her unwavering belief in me, even when I struggled to believe in myself.

**A New Identity, A New Chapter, A New Beginning**

Becoming a Warrant Officer was not the end of my journey—it was a bold new beginning. It prepared me for future assignments. It positioned me for greater responsibility. It opened doors I never imagined. But more importantly, it humbled me. Because I realized then that Jehovah had been guiding my steps long before I knew His name. Every chapter—from Jamaica to Florida, Germany to Fort Polk, Korea to Clovis—was preparing me for this moment. A moment that would lead me toward Kuwait, toward 9/11, toward Iraq, and ultimately… toward faith.

# KUWAIT — THE DEPLOYMENT BEFORE 9/11

When I graduated from Warrant Officer Candidate School, my first duty station as a 140A Air and Missile Defense Systems Integrator was Fort Bliss, Texas, the home of the Air Defense Artillery at the time. I was just starting to settle into the technical rigor of this new chapter when my orders to Kuwait came down. I felt a mix of emotions—pride in my new Warrant Officer role, but a deep ache in my chest knowing I was about to leave Taska and the kids again. It was July 2001. Hot. Heavy. Uncertain. At the time, it was just another deployment. Another rotation. Another opportunity to do my job and represent the Army with excellence. None of us knew that the world was about to change forever.

## Leaving Home Again — The Weight Hits Harder the Second Time

Deploying as an enlisted soldier was one thing. Deploying as a husband, a father of three, and now a Warrant Officer was completely different. The morning, I left, I remember looking at Taska—the quiet strength in her eyes, the softness in her voice. She didn't break down. She didn't panic. She didn't show fear. She held onto me with a strength I can still feel. This time the kids were old enough to understand what was happening. They clung to my legs, wondering if Daddy was coming back this time.

Walking away from them was one of the hardest things I've ever done. Being a soldier requires courage. Being a soldier with a family requires something deeper—a kind of sacrifice that digs into your heart.

### Arriving in Kuwait — Heat, Sand, and a Different Kind of Mission

When I landed in Kuwait, the heat hit me like a wall. Hotter than anything I had ever felt. Hotter than any training environment. Hotter than any place the Army had sent me before. The air felt thick, like you had to push through it just to breathe. The wind carried sand in every direction, and no matter how much you cleaned, it found its way into your boots, your gear, your bed, your mouth. This wasn't Germany. This wasn't Korea. This wasn't Fort Polk. This was a different world.

### A Warrant Officer in a Real-World Mission

As a new Warrant Officer, I carried more responsibility than ever before. I wasn't just another soldier in the formation. I was the technical backbone of the mission: integrating systems, monitoring early warning networks, ensuring communications ran flawlessly, analyzing data, advising commanders, solving problems others didn't see, and keeping the tactical picture synchronized. In Kuwait, real-world missions demanded precision. Mistakes didn't just affect a training exercise—they affected lives, operations, and strategic readiness. Every day required focus. Every night required preparation. It was the kind of environment that sharpened you quickly and humbled you even faster.

## The Rhythm of Deployment

Kuwait had its own rhythm: long shifts, rotating schedules, sandstorms that shut everything down, guarding sensitive equipment, maintenance that never seemed to end, soldiers joking to hide their exhaustion, commanders walking the floor to ask for updates, and briefings that blended into one another. At night, I would sit alone, sometimes thinking of home, picturing Taska handling everything with three babies and no husband to share the load. Her strength pushed me through the hardest days.

## Then... Everything Changed — September 11, 2001

I recall exactly where I was when the news broke. We were going about our day like any other—checking systems, monitoring data, running status updates. Then someone rushed in and said, "A plane hit the World Trade Center". At first, we thought it was an accident. A few minutes later, the second plane hit. The room went silent. Everything around us... stopped. Every soldier looked at the others with the same expression: This is not an accident. This is an attack. And at that moment, the entire atmosphere shifted. It felt like the air got heavier. Like time slowed down. Like every emotion hit all at once: Fear, Anger, Confusion, Determination, Uncertainty, Readiness. We watched the footage from thousands of miles away, but it felt like we were right there.

## The Barracks That Night — Quiet, Heavy, United

That night, Kuwait felt different. No laughter. No usual noise. No casual talk in the hallways. Just quiet. Heavy quiet. Every soldier understood: Our world had changed. Our mission had changed. Our future had changed. Many of us wondered: Would we be sent straight into combat? Was the U.S. going to war? Was our base a target? What was happening back home? Was our family safe? And for me, the biggest question was: How is Taska handling this? Is she scared? Is she watching this alone? Being thousands of miles away from your family during a national tragedy is a feeling you can't fully explain. It's fear and helplessness wrapped into one.

## A New Level of Readiness

Within hours, our base went into high alert. Security tightened. Briefings increased. Missions changed. Leadership shifted focus from routine operations to real-world threat assessments. Given our location, we immediately began preparing for potential retaliatory strikes or border incursions, making the data links and early warning systems I managed our first line of defense. As a Warrant Officer, my responsibilities tripled overnight. Data link architecture, Threat tracking, Communication verification, Radar coordination, Early warning integration, and Support for joint commands. We were no longer training. We were preparing. Kuwait became the staging ground for operations that would be discussed in world history for decades. And I was in the middle of it.

## A Deployment That Became a Turning Point

What began as a normal rotation became one of the most defining seasons of my military life. This deployment: tested my leadership, stretched my skills, challenged my emotions, deepened my sense of duty, strengthened my reliance on Taska, matured me as a man, changed how I viewed the world, and prepared me for Iraq. Looking back, Kuwait was the bridge between the soldier I had been and the leader I was becoming. It was the pause before the storm. It was the moment before everything accelerated. It was the season that pushed me closer to the man Jehovah would one day call back to Him.

# IRAQ — WAR, LOSS, AND THE WEIGHT A SOLDIER CARRIES

War changes you. Not just the missions. Not just the long days. Not just the uniforms, or the heat, or the fear. War changes your heart. War changes your mind. War changes the way you look at life—and death. And nothing prepared me—not the training, not the deployments before—for what Iraq would do to me on the inside.

My deployment to Iraq was not my first time serving in a war zone, as I had previously deployed twice to Operation Enduring Freedom (OEF), once to an undisclosed location and later when I volunteered to go to Germany.

The German deployment was an act of personal sacrifice: A comrade and his wife were both deployed, leaving their children behind. If he went to Germany, he would have immediately deployed again to Iraq. So, I went to Germany knowing that I would immediately be deployed to Iraq in his place, ensuring he could stay home with his family. I was ready to face the war, but I was not prepared for my literal entry into it.

## The Immediate Reality: BIAP Arrival

My first arrival to Iraq was on a C-130 transport plane, and we were greeted immediately by mortars and missiles at the plane. The sounds of battle were instantaneous and overwhelming. Hearing the countermeasures deploy from the plane was a true eye-opener. This was not a soft landing; it was a violent introduction.

Once I got to my unit in BIAP (Baghdad International Airport), the atmosphere shifted immediately. Our unit frequently got bombed; it felt like every hour on the hour. This relentless barrage meant that life was conducted under constant, visible threat.

### Crossing Into Iraq — A Different Kind of Silence

When we crossed from Kuwait into Iraq, the atmosphere shifted immediately. Even the air felt different. Thick. Heavy. Alive with danger. No birds. No normal sounds. Just a quiet you could feel in your bones. Every soldier knew: This is real. This is war. This is not training anymore. We moved with purpose. We checked and rechecked equipment. We scanned every road, every rooftop, every movement in the distance. You didn't relax. You didn't let your guard down. You didn't assume anything. Because in Iraq, anything—and everything—could be a threat.

### IEDs — The Invisible Enemy

IEDs were the most feared weapon we faced daily. You couldn't see them. You couldn't predict them. You couldn't outrun them. Some were buried. Some were hidden in trash piles. Some were placed in dead animals on the roadside. Some were wired into buildings or disguised as harmless objects. Every convoy, every movement, every mission carried the same unspoken tension: "Will it be today?" Even though my technical role centered on airspace defense and communication links, the ground threat meant every trip outside the wire—even for routine maintenance—was a gamble. The IEDs were the constant, indiscriminate reminder of our vulnerability. And sometimes, it was.

## The Night the Radio Broke

While the IEDs were an invisible enemy, one night during my rotation as Battle Captain, that invisible threat became brutally audible over the radio—a sound I will never forget, and one that still hurts to write about. The call came in instantly: an IED had exploded, hitting a truck in the convoy. Through the static, I could hear the Truck Commander (TC) crying out for help, a desperate, raw plea that pierced the command center. Then came the chilling sounds of the gunner and driver. I could hear the blood gurgling in their voices—a sound of life being stolen, live on the network. As the Battle Captain, my job was to be calm, to execute the Immediate Action Drill, and to coordinate the medical evacuation. But as I gave the orders, a part of my soul fractured, forced to listen to the agonizing, final moments of those soldiers, their cries echoing in the silence of the command post. War stops being abstract when you hear the blood gurgling over the radio.

## The Ethical Burden: Looking Down the Barrel

War forces decisions no human should have to make. Amidst the chaos of missions and the constant IED threat, there was one moment that drove the true cost of war deep into my soul. I had to point my weapon at a kid—a child who appeared to be the same age as one of my own. Looking into his face, I didn't see an enemy; I saw a shadow of JP or Tionna, forced into a position of threat. The instant ethical line I had to draw—that agonizing split-second choice between the safety of my unit and the life of a child—was a devastation that no training can prepare you for. This was not about technical expertise or tactical data; it was about the impossible moral price of survival. That image, and the choice I had to be ready to make, is a spiritual scar that will never fade.

## Losing Soldiers — A Pain That Never Leaves

Iraq wasn't just a battlefield. It was a place where men who laughed with you in the morning could be gone by nightfall. We lost soldiers to IEDs. Unexpected. Instant. Devastating. We also lost soldiers to suicide. Quiet battles none of us saw coming until it was too late. Both kinds of loss cut deep—deeper than any wound the battlefield could give. The guilt hits you in different ways: "I should've checked on him." "I should've seen the signs." "Why wasn't it me?" "Why him?" "Did he know how much he meant to us?" War teaches you that strength isn't always loud—and pain isn't always visible.

## The Soldier's Mask — Lead on the Outside, Break on the Inside

As a Warrant Officer, I didn't always have the luxury of showing what I felt. Young soldiers were watching me. NCOs were leaning on me. Officers were depending on me. So, I had to stay calm. Steady. Focused. Even when my heart was heavy. Even when grief hit fast and quiet. Even when I wanted to scream or punch a wall or break down and let the weight out of my chest. But leadership in war demands a different level of strength. A strength that doesn't always feel fair. You encourage others while silently fighting your own battles.

## Long Nights, Unspoken Prayers

There were nights when the generators hummed low, the base quieted down, and the desert wind carried sand across the tents. I would lie awake, thinking about: the soldiers we lost, the soldiers still depending on me, Taska at home, my kids growing up without me there, and whether I would make it home. Sometimes I prayed—not loudly, not formally, but quietly in my heart: "Jehovah... please bring me home." I didn't know Him yet in the way I would later, but I felt Him in those moments—in the fear, in the uncertainty, in the stillness after chaos.

### Daily Life in Iraq — The Rhythm of Survival

Iraq had its own rhythm: briefings at dawn, missions in extreme heat, sandstorms that turned the sky orange, explosions in the distance, radios crackling with urgency, tracking threats, coordinating systems, reporting movement, constant readiness, and constant exhaustion. You learned to live with: lack of sleep, adrenaline swings, fear that felt normal, jokes to hide the stress, cravings for home, and memories you didn't know how to process. But through it all, you kept going. Because stopping wasn't an option. Because lives depended on you. Because quitting wasn't in your DNA.

### Carrying the Weight as a Leader

My role required me to advise commanders, interpret complex tactical data, ensure systems remained synced, prevent communication failures, support air and missile defense coordination, troubleshoot problems under pressure, maintain a clear battlefield picture, mentor junior soldiers, and remain calm in crisis. And beneath all of that, I carried the weight no one saw: the faces of soldiers we lost, the uncertainty of each mission, the prayers for my family, and the hope of making it home. Leadership isn't about rank in war—it's about responsibility. And responsibility gets heavy.

### Moments That Stay With You Forever

There are scenes from Iraq that never leave you: the sound of explosions at night, the smell of burning sand and diesel, the sight of a convoy returning incomplete, the strong silence after bad news, soldiers sitting together without speaking, understanding each other's pain, writing letters home you weren't sure you'd ever send, wearing your weapon like a second skin, and sleeping lightly because danger didn't clock out. War carves memories into your mind and soul. Some you try to remember. Some you try to forget. But all of them become part of who you are.

### The Guilt That Follows a Soldier Home

No one talks about this enough: Even when you make it home, part of you doesn't. You feel guilty for surviving. You feel guilty for laughing again. You feel guilty for normal life. You feel guilty for the soldiers who didn't come home. It's a silent weight that follows many veterans long after their uniforms are hung up. But in my case, that pain would eventually push me toward something deeper. It would become the beginning of my journey toward faith—toward Jehovah—toward understanding purpose in a world filled with pain.

### War Left a Mark — But It Didn't End My Story

Iraq tested me, broke me in places, humbled me, and matured me. I left with: scars you can't see, memories you can't erase, strength I didn't know had, questions I didn't yet know how to answer, a hunger for meaning, a deeper love for my family, and a spiritual emptiness that only Jehovah would one day fill. War didn't define me—but it shaped me. And it prepared the path for everything that came next.

# RETURNING HOME — REBUILDING, REFLECTING, AND FINDING MEANING

Coming home from Iraq is one of the most complicated feelings a soldier can experience. You dream of that moment the entire time you're deployed—seeing your family again, feeling safe again, sleeping in your own bed, hearing your kids' laughter without fear in the background. But when the moment finally comes, you realize something unexpected: You're home... but it takes time for your heart to catch up to your body.

## The Airport Reunion — Joy Mixed with Something Else

When I walked off that plane and saw Taska and the kids waiting for me, my heart felt like it was breaking open and healing at the exact same time. Taska hugged me with a strength that almost made me crumble. It was a mixture of relief, love, gratitude, and suppressed fear that was finally released. My kids ran up to me with smiles so big it almost hurt to look at them. I picked them up, one by one, and felt their arms squeeze around my neck—and for a moment, everything felt perfect. But behind the joy, there was something else: A quiet numbness. A heaviness. A distance inside me that I didn't yet understand.

### Trying to Fit Back Into a Life That Kept Moving Without You

The strange thing about coming home is this: Life doesn't pause while you're gone. It keeps moving. Kids grow. Routines change. Responsibilities shift. Your spouse becomes stronger out of necessity. And when you return, you feel both needed and out of place at the same time. I found myself asking, 'Where do I fit now?' How do I step back into this rhythm? Am I the same husband I was before? Am I the same father? Am I supposed to pretend I'm okay when I'm not sure? No one prepares you for that part. The unpacking of bags is easy—the unpacking of emotions is something else entirely.

### The Silent Weight That Followed Me Home

War leaves you with memories you don't talk about at first. The noise. The fear. The losses. The faces. The long nights. The guilt of surviving. The heaviness that lingers even in quiet rooms. There were nights when I would lie awake beside Taska listening to the quiet breathing of the woman who held everything together while I was gone. And instead of peace, I felt uneasiness. Not because I didn't love my family, but because my mind was still in Iraq. Still listening for explosions. Still expecting danger. Still carrying the weight of soldiers, we lost. You don't just "come home" from those memories. You learn to manage them. To understand them. To grow through them.

### Trying to Be a Husband and Father Again

Rebuilding isn't instant. You try to: reconnect with your wife, reestablish your role, discipline gently, learn your kids again, help around the house, find your place in the routine, and be present. But part of you feels numb. Part of you feels distant. Part of you is trying not to fall apart. Part of you wants to protect them from the things you saw. Part of you wants to pretend nothing happened. And the biggest part of you wants to be whole again. Taska, as always, had a quiet way of giving me space to breathe while still pulling me close enough to remind me that I was home. She didn't push. She didn't pry. She didn't demand. She simply held our family together and let me find my way back one day at a time.

### The Conversations That Changed Everything

Late at night, when the house was quiet and the kids were asleep, Taska and I would sit and talk. Not about the details of war—those were still locked inside me—but about life, purpose, healing, and what came next. She reminded me that I wasn't alone. That I didn't have to carry everything in silence. That the man she loved was still inside me, even if I couldn't feel him yet. She also reminded me that life wasn't just about survival—it was about meaning. Those quiet conversations planted seeds in me. Seeds of reflection. Seeds of purpose. Seeds of faith that would later guide me to Jehovah.

### Seeing Life Differently After War

Once you've been to war, you look at the world differently. Small things matter more: Seeing your kids run in the yard, sitting at the dinner table, waking up without alarms, Sunsets, Quiet mornings, Laughter, and Peace. You stop taking things for granted. You realize how fragile life is. How quickly everything can change. How every moment with family is a blessing. You also become more patient, more grateful, and more aware of how much you need something bigger than yourself. For me, that awareness was the beginning of something spiritual—something

I didn't yet have a name for. But Jehovah was slowly pulling me in, even if I didn't recognize it.

### Rebuilding Myself — One Layer at a Time

Coming home wasn't just about rejoining my family. It was about rebuilding: my emotional health, my identity, my sense of purpose, my leadership, and my peace. I began to reflect more deeply. I began to observe life differently. I began to feel a pull toward something steady and reassuring—something spiritual. Something I had been missing for a long time. Something I didn't yet understand but knew I needed. War broke parts of me I didn't know were fragile. But coming home marked the beginning of the process of putting those pieces back together. Slowly. Quietly. Purposefully.

### A New Beginning — Finding Meaning Again

In the months that followed, I focused on healing: reconnecting with my family, finding my grounding, reflecting on the man I was becoming, searching for deeper meaning, and appreciating every moment I had been given. War had taken something from me. But it also gave me something: Perspective. Humility. Gratitude. A renewed desire to live with purpose. And unknowingly, it prepared my heart for the most transformative chapter of my life—the chapter where I would finally learn about Jehovah and begin to understand who truly carried me through every deployment, every danger, every loss, and every moment I thought I had made it through alone.

# WALKING TOWARD FAITH — HOW GETTING TO KNOW JEHOVAH GAVE MY LIFE NEW MEANING

There are moments in life when you look back and realize Jehovah was guiding you long before you even knew His name. Through childhood lessons. Through challenges. Through blessings. Through war. Through love. Through loss. Through every moment, you thought you were walking alone. For me, the journey toward faith didn't start with a dramatic moment. It started quietly—in the stillness after war when my soul felt empty, and I didn't know why.

**The Quiet Realization: The Space War Left Behind**

Coming home from Iraq left a mark far deeper than I first understood. I had survived things I wasn't supposed to, seen things most people will never have to see, and carried fear, responsibility, and loss in ways that fundamentally changed me. Yet, in the quiet after the homecomings, the reunions, and the noise of "normal life," I realized the war had taken something the peace couldn't give back. There was an emptiness inside me that nothing seemed to fill. Not success, not rank, not accomplishments, not routine—not even the fierce love of my wife and children. I loved them with all my heart, but a part of me still felt deeply restless. I was searching for a meaning that went far deeper than anything I had yet experienced.

## Taska's Example — Love, Faith, and Stability

Throughout our marriage, Taska had always shown strength that didn't come from within herself. Even when she didn't realize it, she demonstrated faith in the little things: her patience, her calmness, her moral compass, her love for people, her ability to forgive, and her unwavering focus on doing what was right, even when life was hard. She wasn't preaching to me. She wasn't pushing anything on me. She was simply living with a heart shaped by faith. And without realizing it, her example planted seeds that would grow later in my life.

## The Early Conversations — Questions That Wouldn't Leave

There came a point after everything I'd been through when I started asking myself: Why am I here? Why did I survive war when others didn't? What am I supposed to do with my life now? How do I heal the parts of me that still hurt? Where does real peace come from? What happens when life ends? Who truly understands our suffering? These weren't questions a soldier or a father could answer alone. And little by little... I began to look upward.

## The Bible Begins to Speak to Me

I was raised within a religious framework, possessing a requisite respect for the divine, yet my understanding remained generalized and imprecise. A profound transformation commenced when I began to acquire an accurate knowledge of Jehovah—a comprehension derived not from mere tradition or transient emotion, but from the verifiable truth of the Scriptures.

A definitive shift in perception occurred. Jehovah was revealed to be neither distant, harsh, nor inaccessible. Instead, the Scriptures illuminated His attributes: patient, loving, forgiving, purposeful, attentive, understanding, and deeply personal. For the first time in my life, the divine being transcended the role of a cultural concept or a family custom. He became tangibly real, demonstrably present, consistently reliable, gracious, and characterized by a gentleness that my heart had not previously encountered.

## Learning Jehovah's Name — A Turning Point

To know an individual's name holds significant power. When I realized that the divine identifier was Jehovah, transcending the generalized titles of 'Lord' or 'God,' a profound shift occurred in my spiritual understanding. The distant, abstract spiritual force dissolved, revealing instead the intimate relationship of a Father, a Protector, and a Guide—an intentional being imbued with identity and love. This discovery of His name fostered an unexpected closeness. It instantly rendered the biblical narrative intensely personal, transformed my faith into a tangible reality, and made the act of prayer deeply authentic.

## Studying the Bible — Piece by Piece, Step by Step

As I began studying, Scriptures that once seemed distant now spoke directly to me. For instance, the promise, "Jehovah is close to the brokenhearted" (Psalm 34:18), resonated deeply, as I had been brokenhearted more than once. The words "I will give you peace, not as the world gives" (John 14:27), held special meaning, since the world had offered only temporary moments of calm. And the counsel to "Throw your burden on Jehovah" (Psalm 55:22) addressed the burdens that had been heavy for years. Bible truth didn't hit me all at once. It grew slowly, steadily—like someone turning up a dim light until the room becomes clear. And piece by piece, I found clarity in places I didn't expect.

### Faith Begins to Change My Marriage and Fatherhood

As I grew spiritually, I also grew as a husband and as a father. I became more patient, more understanding, more present, more intentional, less angry, less guarded, and less controlled by my past. The deeper I learned about Jehovah, the more I softened from the inside out. War had hardened parts of me I didn't know how to heal. Jehovah softened them. Slowly. Gently. With purpose.

### Letting Go of Guilt — A Spiritual Healing

War leaves you with guilt—survivor's guilt, emotional guilt, and the burden of things you didn't control. Yet, learning about Jehovah helped me understand **grace** in a way I had never understood before. I learned that Jehovah sees every tear, understands every trauma, knows every thought, forgives every mistake, values every life, comforts every wound, and gives purpose to every trial. I didn't have to be perfect. I didn't have to hide my pain. I didn't have to be the overly strong soldier every moment of my life. Jehovah allowed me to finally exhale.

### A Spiritual Anchor in a Chaotic World

The world around us continued to change—the military, my career, my children, the pace of life. But through it all, I found something I had never truly possessed: peace. Not because life grew easier, but because Jehovah became my anchor. For the first time, I understood a relationship with God, one built not on fear or tradition, but on love, truth, and meaningful study.

## Jehovah Gave My Life New Meaning

Looking back, I can see that everything—every struggle, every deployment, every moment of loss, every blessing, every challenge—was leading me toward Him. My life wasn't a collection of random events, but a deliberate journey of preparation—to become a better husband, father, leader, and man, and ultimately, to build the relationship with Jehovah that gave purpose to everything I had endured. Faith didn't erase my past. It reframed it. It redeemed it. It gave it purpose. And for the first time, I didn't feel empty—I felt whole. Because Jehovah filled the space war had left behind and gave me peace that the world could never offer.

# EPILOGUE- A LETTER TO MY FAMILY

There are words a man carries inside him for years—words he feels deeply, but doesn't always know how to say. Words that come from the places in his heart that have been shaped by love, by loss, by growth, by experience, and by the quiet moments life gives you when everything finally becomes clear.

**This letter—these final words—are for you.**

**To Taska — My wife, my partner, my blessing:** From the day we met—from that simple moment when you looked at me through the sun visor mirror—my life has never been the same. You were there through every chapter: When I was young and unsure, When I was broke and trying to find my way, When the Army demanded more than I could give, When we became parents quicker than we expected, When deployments stretched on for months, When loneliness and fear weighed heavy, When the world shifted after 9/11, When I walked into Iraq carrying responsibility I never asked for, When I came home with pieces of myself I didn't know how to put back together, and When Jehovah began to reach for me long before I understood His voice. You carried our home. You carried our children. You carried me. Your strength held up everything the Army tried to pull apart. You loved me when I was far away, believed in me when I doubted myself, and built a home that I couldn't wait to return to. I am the man I am today because you stood beside me when life was hard, and stayed with me when lesser love would've walked away.

**To my children — my heart, my purpose, my reason:** Each of you came into my life exactly when I needed you. Your presence changed me. Your innocence grounded me. Your laughter kept my spirit alive in places where happiness was hard to find. Every day overseas, every night in a barracks room, every moment when danger felt too close—I thought about you. I wanted to be a man you could respect, a father you could trust, a presence you could depend on, a protector you could look up to. I missed milestones. I missed the special days at home. I missed moments I cannot get back. But not a day passed that I didn't carry you in my heart. And now that I am home for good, every moment I have with you is priceless. I hope this book helps you understand not just where I was—but who I was, whom I became, and how deeply I love each of you.

**To my family near and far:** Thank you for shaping me. Thank you for believing in me. Thank you for covering us in love during every chapter of our journey.

**To the soldiers I served with:** Some of you are still here. Some of you are gone. Some of you carry invisible wounds. Some of you taught me lessons I will never forget. This book is part of your story, too.

**And above all... to Jehovah:** Thank You for watching over me long before I knew Your name. Thank you for protecting me when danger hid in the dark. Thank you for guiding my steps when I didn't know where I was going. Thank You for bringing me home—not just physically, but emotionally and spiritually. You rebuilt my heart through love, through purpose, through family, through faith.

**To my family:** You are my greatest joy. My greatest lesson. My greatest blessing. My greatest purpose. This book is not about war. It's not about rank. It's not about the Army. It's about you. It's about how your love anchored me through every storm, every deployment, every hardship, every moment I thought I couldn't keep going. I am who I am because of you. And now, finally... after all the years, after all the distance, after all the change, after everything life has taken us through— I'm home. I'm present. I'm whole. I'm grateful. And I am **HOME FOR GOOD.**

# ABOUT THE AUTHOR

**Patrick Plummer** is a retired U.S. Army Warrant Officer who reached the rank of **Chief Warrant Officer Three (CW3)** during a distinguished career that spanned the Cold War era, the establishment of Air Defense command systems, and combat operations in the Middle East. Born to Jamaican parents, he was instilled with a fierce discipline and a drive for excellence that shaped his military life.

Plummer's service began as a **91B Combat Medic** stationed in **Cold War Berlin**, a period that taught him the necessity of raw humility and responsibility. He later transitioned into the complex technical world of Air and Missile Defense, serving as a **14J C4I System Integrator** and ultimately becoming a **140A Air and Missile Defense Systems Integrator**. He achieved the rank of Staff Sergeant during an unaccompanied tour in **Korea** and later served as a recruiter.

His life and leadership were fundamentally tested during his deployment to **Kuwait** and **Iraq**, where he served as a Battle Captain after the shock of 9/11.

**Home for Good** is Patrick Plummer's testament to overcoming the emotional and spiritual scars of war, detailing his journey from the chaos of the battlefield to finding his true purpose, peace, and spiritual home through his relationship with Jehovah.

www.ingramcontent.com/pod-product-compliance
Lightning Source LLC
Chambersburg PA
CBHW050656160426
43194CB00010B/1967